Wounded Warrior with Invisible Scars

A Memoir

By Mataya Wright Williams

Wounded Warriors with Invisible Scars
ISBN: 978-1-7346285-4-8

Printed in the United States of America

Introduction

My Village

They say it takes a village to raise a child; and me, well, I am most grateful for my village. My name is Mataya, also known as Taya to some, and "Black Girl" to others. I was born and raised in Montgomery, Alabama, where I grew up in the best housing project there is, Victor Tulane Court or TLC, as we affectionately called it. I had lots of fun while creating vivid memories and lifetime friendships while living in TLC. The baby of my parents' five children, I enjoyed growing up with my siblings. They considered me to be spoiled; however, they never took ownership of the part they played in spoiling me. I often recall how attentive my siblings were, and how they provided for me countless times in an attempt to decrease the pressure of providing for my parents. They would give me money for the movies, skating rink, salon, and any other things girls my age felt we just had to fashionably have. I did not realize this, nor did I realize how blessed I was to have them until I got older. As I matured, I realized the impact of all the things they contributed to my life. Growing up, my village consisted of my parents, siblings, a few cousins, and my friends.

My brother, Marvin (Prime), was always the big

brother that every girl dreamed of having. He was so funny. I always said he reminded me of the well-known comedian Martin Lawrence when he joked around. As a big brother, though, he was attentive, affectionate, and jovial. I'm certain there were even times he was sure he was Michael Jackson. From the Jheri curl, gloves, and dance moves, you could not convince him of anything different. He would literally turn the dance floor out at any party; it didn't matter whose. Whenever I messed up, he knew how to put me at ease by being cool, supportive, and even-toned. My big brother always made me feel empowered to do anything I put my mind to, which is why I always consult him regarding big decisions. The support and protection he provides me is unlike any other; anyone better not mess with his family, period. Prime is married to Rachel and has two sons, A'travis and Jamarvis, who I treat like my own sons and love them dearly.

My sister, Sophia (Mae-Dudley/Dug), was in the Army by the time I made it to elementary school, but she was, and is, always just a phone call away. I was her first child before she had my niece. Dug always made sure I had school clothes, money for field trips, and anything else I would ask her for when she called home. I loved hearing her stories of life in the Army and learning about the cool places she'd been. Dug was the child that everyone said stayed in her books and didn't hang out a lot. They often said she should have become a lawyer because she was a great debater. She would argue until the end to prove her point. She was always interested in my grades and how well I was doing in school. So, when I told her I planned to go to the Army once I graduated high school, just like she did, she was elated. I don't know if she knew the impact she had on me when she paid for me and mom's first airplane ride to Washington, D.C., in 1989. That was a trip I will never forget. Mom told everybody we were flying to Washington, D.C., but never told them how scared she was to get on the plane. I often refer to Dug as my sister-sister because she is my sister by blood and my sister in arms. She is now married to Gregory Davis, and they have one daughter, Dajah, who is currently serving in the Army as an officer.

My sister, Annie (Baldy), is the outspoken one of the clan. She definitely does not mince her words for anybody under any circumstance. Baldy was the sister who fussed at me, kept me in line, but still somehow ended up letting me have my way. My mom used to joke about Baldy always making sure I was bathed, and my hair was combed. What's funny is, the moment I would go outside, I immediately laid on my stomach and licked the dirt. That didn't matter though; she still spent all of her summer job money on buying me clothes and hair bows despite my love for laying in the dirt. Once I got older, I wanted to wear her clothes and shoes, but she wouldn't let me. She always said I was rough on mine. In hindsight, I'm sure I was. Against her wishes, though, I would sneak and wear her expensive outfits to school but rush home to change before she got off work. Even though I attempted to hide it, she knew I wore things, but I would deny it to the depths. I remember her splurging on me when she used to get her tax refund as if I was her child. Her strength and provision carry on, as she continues to be the sole care provider for my beautiful twenty-year-old niece, Zakayla Chemet, who has cerebral palsy. My sister Baldy is definitely the real warrior in my eyes. During her fight with beating breast cancer, she taught us all the true meaning of faith. I call her sister bear for a reason! She's also one of my protectors—you better not mess with her Taya.

My brother, Andre (Po-Wright), takes the cake when it comes to who spoiled me most. He always went overboard and gave me the best of anything and everything I wanted or needed. Like Baldy, he kept my wardrobe on point and always made sure my shoe game was unmatched. In his eyes, there was nothing too good for his baby sister, and he had no problem showing it. We have an incredible bond. Being the babies of the family, he would let me get away with just about anything. Po-Wright was very protective of my future, so he was sure to keep me focused on the right things. He always told me, "Don't you look for no man to give you anything in these streets, I got you," and he certainly did! His friends were like brothers to me growing up. They were always on the lookout for me whenever he wasn't around. It was a gift and a curse; a curse because I couldn't get away with anything! My brother is also known as the family party starter. He sang and danced anywhere, and Lord, don't let a microphone be anywhere near because he is going to sing until the lights come on or the last call for alcohol. Po-Wright was well respected in the community with the kids because he devoted a lot of time with young boys coaching little league sports or the local schools. He now has four sons of his own, Rodriquez, DeAndre, Courtland, and Andre Jr., and one daughter, Zykerria, whom I love dearly.

Much love and admiration to my extended brothers and sister: Willie Scott, Jr., Greg Hudson, and the one and only Carolyn "CB" Billups. We had different moms but shared the same dad. My mother loved them as if they were her own. We have so much love between all of us, even with me being the baby of all. Being the baby, I get to hear the funny stories from everyone else, stories about how things were for them all growing up. One thing I do know is that dad spoiled all my siblings equally and continued the trend when I came along. I was my sister Carolyn's baby sister for real, and she never missed an opportunity to let the world know it. She was the only one that could convince our dad to let me do things he wouldn't normally allow me to do. He could tell me, no, but once Carolyn asked, eventually, his answer would be yes. I looked forward to spending the night at her house on the weekends because I knew we were going to have some fun. My niece, LaShunda Billups, and nephew, Tommy "Blvd" Billups Sr., always showed me a good time when I was at the house with them.

All the stories I heard about my older brother Willie Jr. are consistent with what people in Montgomery typically said. He was known as the super cool, fly, nicely dressed young brother riding around the streets of Montgomery with Stacy Adams on. I'm pretty sure he invented the "swag" before it became a thing in our culture today. I always went to Atlanta in the summertime to hang with my nephews, Darrell "Bam" Carter and Christopher "Lips" Carter, and niece, Rotonia Carter Wright. My niece and nephews seemed to know everyone from College Park, Riverdale, Decatur, and anything in between. When I was in Atlanta, I was everybody's auntie, and eventually, all their friends started calling me the "Lil Auntie," because my niece, nephews, and all their friends were a lot older than me. That didn't matter though, because where they went, I was right there with them, always on the scene. Their mom, Linda, told them they had to take me with them, and they had no problem doing just that. Hanging out with my older niece and nephews taught me a lot and made our bond closer too. What I think solidified our bond was the fact that I never told my dad what we did when we were out and about in the streets of Atlanta, and from then on, they had no problem letting me tag along.

My brother Greg and I were closer in age than me, Carolyn, and Willie Jr. Our closeness in age made it easier for us to keep one another in trouble. For some reason, though, dad never fussed at us when we got into things we weren't supposed to get into together. Because of dad's leniency, we kept getting into things we had no business until we got a little older. Greg and I still have a great relationship and never go too long without checking in on one other.

My brother-in-love Greg Davis, Dug's husband, typically encouraged me and lent support in times of need as well. He never treated me like an in-law, but more like his blood sister. I could talk to him just like I could my sisters and brothers, and it would remain between us. He always gave me sound advice and took a special interest in whatever I had going on. Greg accepted the fact that I was growing up way before my sisters and brothers accepted it, and sometimes had to remind them I was no longer the little girl they had to take care of. No matter what the situation was, he was always front and center in support, like a proud big brother, cheering me on. Greg has definitely been an integral part of my village, and I am extremely grateful for him. Even now, he is sharing that same wisdom and knowledge with my son, who looks up to and loves his uncle Greg.

I had several male cousins and a few others who looked out and kept me focused as well. Many times, we forget about the impact others have had on you that contributed to the good in your life. I remember being out and about, and I would get lectures from my cousins Zavious (Chin Woo), LaHorace (BBop), Eric (Jimmy Lee), Brian (Stripe), or George (Main Main), and my dear friend Rodriquez (Big Pooh). They saw potential in me and wanted to keep me focused. I always looked at it as they were fussing again, but in reality, they were bringing out the best in me. They told me why I shouldn't go to certain places or hang out with certain people. I didn't immediately appreciate their guidance. I just wanted to be a teen and enjoy life. I didn't quite understand their methods until I matured a little more. Every year for Valentine's Day, Jimmy Lee would buy me candy and balloons. Back then, he probably didn't realize the impact the gesture made in my life, but it made a lasting one. Many times, girls my age would be swept off their feet if a guy, young or old, gifted them anything. My cousin Jimmy Lee made me feel genuinely loved and reminded me I didn't have to look for love in the streets because my family loved me more than enough. He was my first true valentine.

My senior year in high school, our bus, which was zoned for Jefferson Davis High School, was suspended, making it the parents' responsibility to provide transportation if they wanted their children to attend the school. Although I was a senior, I didn't have a car and honestly couldn't even drive, but my dear friend Big Pooh committed himself to getting me to and from school every day. He never charged me anything or expected anything in return. I believe he kept his commitment because he wanted me to graduate just as badly as I did. I didn't want to change schools, especially because it seemed to me as though the school system was intentionally grouping us all together in a way that set us up for failure. They put most of the neighborhood kids who had differences in the same school, which limited the overall diversity but increased the chances for problems and failure to occur. Big Pooh knew my desire and had no problem stepping up to help me succeed. Rest in heaven, my dear friend, you are truly missed! My lovely cousin, Tiffany Hall Shaver, always took the time to make sure I was up on my studies. She became my personal study buddy from junior high until graduation. She vowed never to leave me behind, and she meant that. I remember vividly times where we studied for tests in her mom's kitchen, worked countless projects, and massive reports. Tiffany always challenged me and motivated me to be my absolute best at whatever I did.

At the end of each day, family and friends are so important; and having love surround you make any difficult situation bearable. This is my village.

Taking the Oath

Summertime was always my favorite time of year because my birthday is in July. The year I turned seventeen, I couldn't wait! Apparently, my Army recruiter was just as excited because one of my first "happy birthday" phone calls came from him. I was an easy sell because by this time I already knew I wanted to join the Army like my big sister, Dug. For as long as I could remember, I saw my future self traveling the world, meeting new people, and experiencing new things. I never had aspirations of becoming a doctor, lawyer, or teacher, which were typically the career topics of discussion among girls and other classmates my age. Upon making my decision to enlist in the military, I received input from both ends of the spectrum. I received encouragement from many people, and on the other hand, many people told me that I should not go. My mom and dad were not so happy about my decision to join the military at all, but they never once attempted to discourage me from making the choice to enlist in August 1997. I expressed to them I wanted to see the world and told them I knew there was more to my life than what I was exposed to. All things considered, my parents signed the dotted line in support of my choice of joining the military.

The start of my senior year was exciting, and I was happy to finally be a senior walking the halls of Jefferson Davis High School. I was so proud to make it to that point, and I wasn't going to let anything stop me from graduating. I struggled terribly with the graduation exit exams. This was primarily because I wasn't aware of how important they were; no one ever made that clear to me until my senior year. After I received my first round of exam results, my cousin Chin-Woo asked how I had done. I shrugged my shoulders nonchalantly and told him, "I only passed one," which was an English exam. He said to me, "Cuz, you gotta pass them all before you can graduate," and in disbelief and awe, I responded, "Huh?" I immediately buckled down and focused on getting myself ready for the next scheduled test.

During this same time, I was introduced by my male best friend, Shang, to a guy named Courtney "Luck" Williams. After we were introduced, I immediately and proudly told him I was going to the Army after graduation. Even though I made it clear I wouldn't be in Montgomery much longer, we hit it off good. Instead of attempting to discourage me from leaving for the Army, he encouraged me and even helped me to stay focused on my goal. There was something different about him; even though we were young, I could just tell there was something special about him. After being acquainted for a while, he finally met my mom. After their first meeting, she did not take her time telling me Luck would be my husband. My mother was crazy about him, and I kept reminding her, "Ma, remember I am leaving this summer, and he will probably find a new girl and forget about me." I was totally wrong! Graduation came May 1998, and despite the fact that I would soon be shipping out to Fort Leonard Wood, Missouri, that July to start my basic training, Luck and I stayed in a committed relationship.

When I arrived at basic military training (BMT) in Fort Leonard Wood, I still had the attitude of Taya from the TLC projects, and a gold tooth to add. That alone made me an easy target for the drill sergeants to have fun with pointing me out and picking on me. I was challenged in ways unimaginable and many times had to dig way deep and push myself to the limits. I was in an infantry unit for training, which during that time wasn't accustomed to having females integrated, which meant there was no distinction between male and female. Boy was I in for a rude awakening. By week three, I felt like my drill sergeant hated me. At every turn, I heard Private Wright, drop and give me fifty push-ups. Private Wright, run and show me far (run from the farthest place from him at full speed and back to him without stopping). Private Wright, get down and do some flutter kicks until you get tired. Private Wright, drop down and beat your face. And the famous rhetorical question, are you looking at me Private?! Get down and get some. That was torture on top of the 0330 formations for roll call and physical training, miles long ruck marches, M16 rifle qualifications, hours of classroom lectures, field training exercises, blistered feet, hair loss, and dehydration. I will never ever forget that experience. There were many times when I would call Luck complaining about some of the crappy stuff I had to go through, and he would encourage me as usual. He would always tell me I was made for the challenges, to stay focused on my goal, and just remember why I was there.

After graduating from basic training, I was sent to Fort Lee, Virginia, for advanced individual training (AIT), where we were trained specifically for our military occupational specialty (MOS). We all showed up to our first day of class nervous as ever, not knowing what to expect of the challenges that were ahead. The course was off to a great start, and we were all focused and eager to learn as much as we could from the staff sergeant instructors. Unfortunately, the only thing the three instructors were trying to show us females was what was in between their legs. They all flirted with the African American female soldiers every chance they got. They encouraged us to break the curfew rules, sent cabs to our dorms, and even though we never went, continued to send them. They offered alcohol to us while we were still in training status and legally underage to even drink. I could tell it was a pattern of theirs; we all did, but no one ever told anyone because we didn't know how the process worked.

One of my instructors was bold enough to ask me if I would call him once I got my off base overnight privileges. I looked at him in disgust and walked away; he was old enough to be my father. He later told the other instructors I had an attitude and wouldn't "cooperate." From then on, all the instructors began giving me a hard time and tried making the course a lot more stressful for me. Thanks to who I was, and the support system I grew up having, I was mentally and emotionally strong enough to withstand the conditions in which I had to train and passed with high marks. I was so ready to leave that place and put all of that behind me. Graduation day finally arrived, and I couldn't pack my bags fast enough. I didn't want to stay in that place another day, so I threw away many good items that I wanted in an effort to pack quick, fast, and in a hurry. My dad and niece Rotonia came to my graduation, and I was so happy to see them. I couldn't even begin to tell them of all the things I endured while there. So, I didn't. I just sucked it up, enjoyed their presence, and eagerly anticipated my arrival at Fort Hood, Texas, because I was officially a soldier.

You Can't Trust Everybody

I arrived at Fort Hood, Texas, as a super nervous eighteen-year-old private second class (PV2)/E-2 and didn't quite know what to expect. I had my boots shined, and uniform pressed to the nines. I met my sponsor, who picked me up from the airport wearing a big smile as he shook my hand. He said I would be assigned to him during my transition into the new unit, and he would be looking out for me. Grateful, I thanked him because I had never been on my own or even away from my parents for long, so having a sponsor or someone assigned to support me along that portion of my journey was appreciated. He assured me not to worry stating, "I got you," and told me he would also show me around in the process. I trusted that my sponsor had my best interest at heart because I didn't have any other reasons to think he didn't. My initial thoughts were that the Army surely knows how to take care of new soldiers. All along, this particular soldier had other plans.

As I got settled into my new post, I was made more and more comfortable throughout the first few days, and I even started to trust this African American brother. At this point, I felt safe and secure. I called my mom and dad and told them I had made it safe and sound and bragged about how I had a sponsor to meet me at the airport. I did my initial meet and greet with my section supervisor and other soldiers from my work section in the motor pool. My non-commissioned officer in charge (NCOIC) was cool, and he definitely made me feel welcomed as well. I still didn't have a car because I still barely knew how to drive, so I depended on the one person who assured me I could trust him to look out for me. Yep! My sponsor/squad leader. He took me grocery shopping, to the post exchange, and other places I needed to get to since I wasn't familiar with the area.

During this time, I wasn't making enough money to buy a TV, so I read a lot of books, magazines, and tried to make friends with the soldiers in my barracks. Being that it was mid-January, many soldiers, including myself, were in their rooms due to the weather. I occasionally saw some soldiers in the laundry room or the day room, but not very often. One day, my sponsor showed up at my dorm and told me to stop sitting in my dorm room. He said, "A few folks from our section are stopping by my crib later this evening to eat because I threw down in the kitchen." He reminded me I told him I liked to read and said, "I have a ton of books you can grab off my bookshelf while you're there."

I decided to take him up on that offer to eat a good home-cooked meal, plus I didn't have a female friend just yet and anticipated connecting with the ones from my section during the social gathering. We arrived at his place, and the food was already done and smelled delicious. He put a movie on while assuring me the others should be arriving soon. Every movie choice was pretty new to me, considering I had been in training for the past six months, so I had to get caught up with life outside of the military all over again. I even had the opportunity to take a good look at his bookshelf, which is where I first learned of the author Eric Jerome Dickey. During my time of movie watching and book browsing, I kept asking about the other folks from my section showing up, but never got a clear answer on their arrival time. Shortly after me inquiring a few times, he began making excuses for their tardiness and eventually said he didn't think any of them would make it. At the time, it didn't bother me much, I was good, I trusted him. I made myself more comfortable with the idea, started reading one of the books I was borrowing, and felt safe. As he noticed me getting more comfortable, he started making inappropriate comments. I picked up on that quickly, though, and started hinting that I was ready to get back to the barracks to get ready for work the next day.

It was a Sunday evening, and I still remember it vividly like yesterday, even though it was over twenty years ago. As we continued to talk, he got closer and closer to me before finally leaned in to kiss me. In complete shock, I pushed him back away from me as I jumped up, demanding to be taken back to my room. He began telling me I owed him for all he had done for me, saying how ungrateful I was and that he knew I wanted it. Even in my eighteen-year-old mind, I knew it was going to get real bad, real fast. He grabbed me and threw me down on the sofa, and literally tried to take my clothes off. I fought, kicked, screamed, and got out of there! I didn't know where to go, how to get back to my room, and definitely didn't have anyone to call, but I ran until I felt like I was far enough away from him. I was crying and running down back streets as I tried to remember the route we took to get to his apartment. The car ride wasn't that long outside the post, so I knew I could make it. I knew in that moment God kept me because even through my tears and fear while running, I looked up and saw the Fourth Infantry Division water tower and continued running towards it. I finally made it back to the post, but I didn't have to show my military ID card. Luckily for me, Fort Hood was considered an "open post" at the time. I remembered my barracks were off 37th street. Still running, looking at the street signs, I finally made it to my building and found my key. It was only until that moment I felt genuinely safe. I couldn't fall asleep that night because my mind was trying to process what had just happened. Later, there was a knock on my door, and I just laid in bed quiet and still. I figured it was my sponsor, but I refused to let him know I was in there. I laid still until I fell asleep out of fear and exhaustion from all that had happened.

I finally woke up and was anxious to get to work. I didn't even know where to begin with letting anyone know what had happened to me. What I did know was that it felt good to be back at work around other people. I finally worked up the nerve to explain what happened to one of the sergeant first class, but unfortunately, he brushed it off as he told me the sergeant that I was "accusing" was a good guy. He went on to assure me he worked with him for a long time, and accusations like that could ruin a good person's career. The most he offered to do was to talk to him about it. I'm not sure how, but soon after, word began floating around the work center that I'd had sex with the sergeant. I'd been approached by male and female soldiers saying they heard I took everything so seriously, and I needed just to relax and lighten up.

That was my first experience with the broken system put into place to protect us. Soldiers and NCOs were more conditioned to dismiss the actions of the offending soldier, protect them, and victim blame. Shortly after that encounter, I learned I wasn't his first. I was just a blessed young girl God protected at that moment, not to become another one of his victims. My sponsor even had the audacity to tell other guys in casual conversations he had sex with me! That was not only embarrassing, but he was openly demeaning and defaming my character without a care in the world. He was even bold enough to harass me at work, so I would try to avoid him at all costs. To make matters worse, because he was still my supervisor, he found things on the job to get on my case about and often locked me at the position of attention while screaming in my face. I remembered standing at the back dock of our building one day as he chewed me out, and tears streamed down my face. I asked him several times why he was doing that to me and why he lied about having sex with me. I think that gave him a rise as he never really gave me an answer, just further unwarranted retaliation on the job.

As a sergeant, he controlled my performance appraisals and could recommend or deny my promotions. I was only an E-2 at the time, so I genuinely felt like I wasn't going to survive him. Until one day, I found my voice and wasn't afraid of the consequences of being written up or whatever reprimand that came from me speaking up. I opened my mouth, and the words literally just rolled off my tongue. It was like the room stood still, and all eyes were on us from everyone in the room. He tried to silence meby using his rank and over talk me. I blurted out to everyone he tried to force himself upon me and have sex with me. I told them how I escaped his apartment and how he was using his rank against me to control me because I got away, and I didn't want to be around him. I never had a problem with him after outing him to our peers. He eventually went to another duty assignment, and I honestly didn't care if I ever saw him again. I resented my leadership for taking up for him and not holding him accountable for what he tried to do to me and had gotten away with doing to so many others before me.

As I matured in my career, I told others about my experience, and they told me what I should have done. I often responded with a simple, "I should have been given this information a long time ago." I felt so defeated, and all things considered, I really looked forward to the end of my service contract. Instead of voluntarily renewing or getting out altogether, I ended up going to the Army Reserves due to Stop-Loss. This was shortly after the terrorist attacks on September 11, 2001. Military Stop-Loss is an involuntary extension to active duty members that kept you beyond your contract term for mission purposes. Everything was a whirlwind as training ramped up, and soldiers were headed out to fight the War on Terror daily. I was released from Fort Hood immediately after Stop-Loss was lifted.

My Next Round with the Army

I arrived at my Army Reserves unit in Alabama, which was activated to go to war to assist with the War on Terror shortly after I arrived. As we prepared to go to combat, there were underlying internal wars within my new organization, which consisted of racial and sexist tensions. It was immediately noted that African American soldiers didn't have equal opportunities for promotions. As I looked around, every person in leadership was a white male, all upward mobility opportunities were pre-selected for white males, and the leadership would move unqualified white males in positions that black males/females qualified for but weren't promoted to. The leadership would promote the white males to the highest rank in the section and use the African Americans to do the work.

We were mobilized from Fort Stewart, Georgia, in preparation for deployment; I didn't know just how bad things were until they happened to me. We had a group of good ole boys who were all white males that looked out for their own. Meanwhile, black service members were treated as second class citizens. As we approached our shipping date to land in Iraq, my leadership informed me they made a decision to leave me back to deploy with another unit. A unit in which I knew no one, using my slot with my initial unit for a white male coming from a Kentucky unit. Despite how I felt about deploying with a "stranger unit," I was immediately notified the decision had already been made, and it was final. I asked questions regarding the decision, and no one seemed to care about my concern. As a female, I was concerned I would be reassigned with a group of total strangers, and I was not completely comfortable with that. Especially given the history I'd had in the past. The only answer I received was they didn't have to explain why.

I solicited help from everyone who was in a position to help me and was shut down at every angle. I went to my master sergeant, who was a black male, and he literally said to me it was nothing he could do to help me because "You know those white boys are going to do whatever they want to do." I was floored! Especially since he agreed they were wrong but wanted to save himself instead of saving me. I just wanted someone to stand up for what was right. I'd never seen so many people who had a voice but remained silent and wouldn't fight for what they knew was an absolute injustice. I had so many questions I just couldn't seem to get the answers to. No one even cared enough to pretend to care about the situation I was in. No one knew which unit I would be reassigned to and didn't care because if things had worked as they planned, they would be long gone before I was picked up and shipped out. I cried so much and prayed to God and often stated, "Lord, you see what I am going through with these folks, and they can't keep getting away with the stuff they do." I was so frustrated and lost faith in the system once again. At that time, I was a specialist (SPC)/E-4 and learned a lot more about the military, because I wasn't going to continue to suffer at the hands of toxic leadership.

I continued to pray and spoke to my brother Prime, who lived in Georgia, about what I was going through numerous times. He always said, "Baby sis, just stay here in Georgia with me and don't go." His wife was currently deployed, and he and my nephews got used to me being at the house with them during my downtime. I told him it wasn't that simple; they were transferring me to another unit, not leaving me, and I felt they were serving me an injustice. I was mobilized in my billet and tasked to do my job with the people I came with, and that was what I was going to do. I knew that God had finally answered my prayers when a sergeant major on duty was walking around checking the wellness of deploying soldiers. Many soldiers were complaining about the rundown and filthy barracks we were living in that were literally lead-based, and the crappy deployment process as a whole. The sergeant major finally made his way around and approached the group of soldiers I was with and asked if everything was going well for us. I was hesitant to speak because of my past experience with "help." I finally spoke up and told him what was happening as far as my deployment slot being given to someone else from Kentucky, and the plan for me to deploy with the "stranger unit." I explained how I was informed by my leadership they were using my position for someone unqualified, and they were going to deploy me with a group of people who I didn't even know, nor did I know the mission they were assigned. I told him I was trained to do my job and not a job they were just going to put me in. He listened, he asked me what seemed like a thousand questions, he approached the leadership, and at the end of the day, I was rightfully put back into my position and went to war with the unit I was mobilized with. My leadership didn't take too kindly to this move by a little ole' E-4. They wanted to know who I thought I was, and I replied, "a soldier".

Mosul, Iraq 2003

The War on Terror was ramping up, and my unit and I were immediately thrust into battle. While our Iraqi battle was underway, the battle of racism within our unit was getting worse as well. We finally landed in the country and started our in-processing. During in-processing, we were getting all sorts of classified briefings, rules of engagement, and maps of the logistic support areas (LSA) that we would stop at on our days-long convoy to reach Mosul. After in-processing, we headed up north, and it was a horrible time for American soldiers. We traveled through some dangerous cities like Fallujah, Tikrit, and Baghdad, just to name a few. We had to literally fight the locals, just to get through the streets in our high mobility multipurpose wheeled vehicles (HMMWV). The rules were simple, keep driving, and never stop until we reached our destination. The things we saw along our journey were a sad sight. It almost felt like we were bullies, but it was the only way we would survive. As we passed through various cities, our vehicles were stripped of anything not tied down or locked because the locals wanted food, water, and American money due to its value. We saw children marked as unified bombers. Unfortunately, though they were children, we had to train our minds to see killers because they were no longer babies and seen as a terrorist threat. The bombs on their back would potentially take me and my entire unit out, killing themselves in the process. They were willing to do anything for Allah, including kill and die, while others screamed out because they felt we had no right to be in their country. Despite the violent threats and raiding of our vehicles, some natives received us well and saw us as help. Regardless, whether they trusted us or not, we could not trust any of them during this mission. After days of convoying through the dangerous roads, we finally made it to Mosul. By this time, both wars were ramping up—the one with the local enemy and the one with the enemy within.

When I called home, I often told my mom about how many battles we were fighting, and she would say, "Just hang in there, baby. You are strong, and you can handle it." I remember speaking to my dad as well and telling him about some of the things that were going on, and he would be extremely tight-lipped. Those were times I knew for sure his silence meant he was furious. When he did speak, he only wanted to know if I was okay! He would also say, "When you make it back here, I need you to point those bastards out to me, do you hear me?" I would simply say, "Yes, Daddy, I hear you." My dad often told me, "We didn't let you sign up to be treated like shit either," and told me he was sorry I had to deal with the things I dealt with on a daily basis. My dad was so protective that he also wanted to speak to my immediate supervisor. I told him that wasn't a good idea because I still had to be there with them for however long we had left. He remembered meeting my supervisor before we left Alabama, and he assured my dad he was going to take care of me.

I recall sitting at the computer with one of the female soldiers I hung out with, using the internet, trying to send e-mails to our families back home. Apparently, her male sergeant called her name, but she didn't hear him. He then rushed over to where we were sitting and attempted to push her out of her seat, saying she needed to answer him when he called her name. In bewilderment, we looked at each other while the sergeant kept attempting forcefully to push her out of her chair. She was holding on to the chair for dear life, and during this altercation, I was saying to myself, "What in the hell is going on?" The sergeant kept screaming for her to stand to her feet after consistently and unsuccessfully trying to push her out of her seat. The soldier refused by stating, "You don't have a right to put your hands on me!" While this is happening, I was screaming at him to stop and asking the other soldiers in the room to stop him while she is crying and asking him to stop touching her. He replied in a cocky tone, "I can touch you whenever I get ready, now stand the fuck up before you regret it!" No one came to her rescue or tried to intervene. We reported the incident immediately after it happened to the first sergeant. The female soldier and I had to write statements of the event and tell the same story over and over again, getting our hopes up thinking there was no way he would get away with that. Unfortunately, nothing was done to him because he was a white male, and she was a black female. All the witnesses were white males with rank who pretended and verbally stated they didn't see anything. Once the entire investigation process was complete, and nothing was done to him, I was literally sick to my stomach. The system had failed me yet again.

There was another time where I saw a different white male sergeant slap a black male specialist. This happened because the sergeant said the specialist disrespected him. For this incident, there was another investigation where they went through witnesses and written statements, but again, nothing was done. During this investigation, though, leadership only seemed to be concerned with the black male becoming disrespectful to the sergeant during the verbal altercation. The specialist was given a letter of counseling and had his weapon removed, because the leadership feared retaliation. The sergeant received no punishment for his actions.

The first sergeant we had at the time of these events was afraid to speak up because he didn't want to lose his position. He was an African American master sergeant who often turned a blind eye to things that were not right just to stay in the good graces with the all-white male leadership. When I spoke to him about what was going on, he often told me I needed to learn how to play the game. My response was simple, "Playing the game shouldn't consist of selling out." Unfortunately, the first sergeant was looking to make sergeant major and wasn't going to let anything stand in his way, even if it meant doing the right thing by calling the leadership on their bullshit.

We were divided on every front, and the leadership was the driving force of the division. You couldn't make a complaint against them to them, so for many months, we suffered. I became angry, defensive, withdrawn, and focused on my survival on and off the compound. Our battalion commander, while deployed, was attached to us from an Army Reserves unit from Florida. He was simply there to fill a slot and didn't get involved much with the Alabama leadership. He really didn't care what happened one way or the other. He simply let them do what they wanted to do and believed everything leadership told him. As long as the commander didn't have to make a command decision, he was having a good day. The leadership would have their conversations about how stupid he was, so whatever he wanted to do that kept him away from them, they encouraged it.

At one point, the commander decided to create a convoy team called the G-Force, simply put in place to escort him around the northern part of Iraq. This team was strictly put in place for his leisure of galivanting, visiting Saddam's palaces for photos for his family, and bragging rights amongst his peers. Filling every seat in the vehicles in this convoy were the African American soldiers who were the makeup of this illegal travel team. The leadership over transportation intentionally assigned African American soldiers only, which made us feel as though we were expendable. We often came up with excuses for why we couldn't go, but we were told that the G-Force was mandatory, so there was no getting out of it. The non-mission related escort regularly took priority over our war-related duties. We were in real-life combat and lost soldiers everyday mostly to roadside bombing and small arms fire during these escorts. Although the rest of the chain of command didn't agree with the convoys, they didn't care enough to speak up because it was "us" who made the team and put our lives on the line daily. As long as it was us putting our lives on the line and not theirs, they weren't speaking up for us. Members contracted leishmaniasis due to the filthy places we would sleep. Leishmaniasis is a parasitic flesh-eating disease carried by sandflies and is very common in that area. Soldiers were sent to Walter

Reed in D.C. for treatment but sent back to war while we assisted with their wound care and medication. Even with all this happening, they never stopped the illegitimate convoys.

In July 2003, things got even more real when we witnessed a three-hour firefight right outside our compound between U.S. Forces and Saddam Hussein's sons Uday and Qusay. That was one of the most terrifying moments during the deployment to witness something so threatening so close to our place of refuge. All we thought about was trying to stay alive. In December 2003, Saddam was captured by U.S. Forces and believe it or not, even though Saddam treated many of the Iraqis like trash, they were loyal to him. The cities surrounding us were in an uproar, and you could hear them screaming, howling, burning down buildings, and shooting all throughout our area because their leader had been captured. The Iraqis started attacking more and consistently shot rocket-propelled grenades at our compound. The death toll began to rise exponentially, with American soldiers being killed in retaliation.

I remember doing headcount for midnight chow, and a group of soldiers heading out on a mission came through the chow hall. We exchanged pleasantries and talked a little bit about the mission they were heading out on. The last words I remember saying to all of them were, "Be safe out there."

Unfortunately, one of my friends, Specialist Wolf, didn't make it back due to a roadside bomb that detonated on their route. I was informed by members of his team he was killed instantly by the shrapnel during the explosion. He was exposed to elements due to his job as the saw gunner, manning the weapon gun mount. This happens when a vehicle rolls over a roadside bomb, and when it detonates, fragments of a bomb, shell, or object are projected out by the explosion. Shrapnel has the ability to slice through human flesh, causing awful and deadly wounds.

Our combat mission still required us to perform our usual convoys to different worksites, and bank runs to withdraw money for the many projects of rebuilding we were in charge of. The unit was in charge of the combat-heavy engineering piece of this particular phase of our mission. We escorted the contractors every Tuesday to withdraw a million dollars for projects and other outrageous government expenses. We ate at the local restaurants from time to time and even had a chance to view the Tigris river up close. I was eventually tasked to be the non-commissioned officer in charge (NCOIC) of the laundry facility. I supervised a total of thirteen local nationals who owned their own businesses. I was assigned a translator who spoke fluent English and would communicate with them for me. At the time, I could only speak minimal Arabic and needed his help in a major way. They were in charge of washing and drying all the soldiers' uniforms on the compound and were paid through government contracts. Many of them expressed they were making more money than they had ever made in their lives by working for us.

Even with all of this going on, the African American soldiers were still subjected to racial slurs, physical assaults, bathing outside, literally using jet fuel and fire to burn our poop, pee, and trash, no running potable/water, and still fighting Iraqis off the compound. You have to ask yourself at some point, how much could one take? At that point, I viewed my Army career as finished. I spoke it over with Luck and decided I was getting out. I still couldn't believe the way things existed in the military. I also spoke to my company commander as my enlistment was coming to a close. He suggested I give it another try because I had so much to offer the military, and I shouldn't give up. I had joined for a purpose and decided not to allow the actions of others to determine my future. I was committed to my service, and I loved being a part of something that was bigger than myself—even though that something couldn't figure out how to quite treat me with the respect I deserved.

My Boaz

In 1997, back in high school, I was introduced to Courtney "Luck" Williams, Sr. by his cousin Shang Thomas. Shang was one of my high school best friends, who I talked about everything with. I distinctly remember a time after a breakup where I shared with Shang I was not interested in a boyfriend at the time. I told him I was just going to focus on graduation and head to the Army. Shang said to me, "Black, I got somebody for you." I looked at him and said, "No, you don't!" Shang immediately told me how nice this guy was and assured me he was my type. A few weeks had gone by, and he finally introduced me to the guy he had deemed to be my type. I wasn't unfamiliar with the guy because I had noticed him in school between classes in the hallways. I thought he was really cute, but again, I was focused on heading to the Army and not having a boyfriend, and that was that. We held a small conversation and exchanged numbers. I must admit, boy was he different. Without delay, I shared with him my plans to leave for the Army, and I told him my date for basic training was already set for me to ship out after graduation.

Luck and I kept in touch while I was away in training, and he often visited me at my stateside duty stations. He was one of the main people who kept me grounded amidst all the things I had endured during my time of service in the military. My mom loved Luck dearly from the start, and she was beyond convinced he was going to be her son-in-law. She was absolutely sure of it. On July 30, 2004, my mother's prophecy manifested. Luck and I made our union official to begin our journey toward Williams forever after. We didn't have a traditional wedding which is typical for military families. We eloped at the courthouse in our hometown of Montgomery, Alabama, where my parents, my sister Baldy, his grandmother Rose, and sister Sharn were there as our witnesses and support system. This day was one of the happiest days of my life.

Luck is, by far, a blessing from God. He has been my rock as far back as I can remember. He has supported me through the highs, lows, and extremely rock bottoms. He has been a solid support system throughout my military career, believing in me when I didn't believe in myself. He always made it his business to make me study, quiz me for tests, and just keep me focused. No matter how far Uncle Sam sent me his love, support, and affection was always what got me through. He is so attentive! We became best friends before anything, and that is why our connection is so strong. He loves me for me, and I never had to pretend to be anyone other than myself. With him, I am respected, protected, and loved. It is so easy to love someone who loves you just as much as you love them. We have grown so much together over the past twenty-three years, and God is continually blessing us. Luck has made many sacrifices along this journey that I am more than grateful for.

There were times where he took vacation time from work to visit me wherever I was stationed. We explored the different locations and experienced new things together. We did things like explored landmarks, visited parks and museums, experienced a variety of foods, and of course, did our fair share of shopping. By traveling the world together, we created many memories that will last us a lifetime while intentionally making time to create many more. On October 17, 2006, we were blessed to have our son Courtney, Jr., affectionately known as Dooney. Together, we have committed to being the best parents and examples we can be. Our hope is to raise a son who will defeat the odds and not have to deal with the many things we did growing up. Luck is by far the BEST dad to our son and the best in-house example of what a man is. They have an incredible bond, and Dooney looks up to his father. I admire the talks Luck has with our son, and the explanations he gives him on why/how to do certain things; things that, as a mother, I could never teach him. Our family time is amazing and competitive because they both go out of their way to have bragging rights for whatever challenging game we decide to play. Dooney and I often talk about the great things his dad brings to our family and how we are blessed to have him. He literally tries to teach him everything!

I can remember being gone on deployment and returning home jet-lagged, and Dooney was anxious to show me this "new" thing he learned while I was away. I woke up from a nap hearing my then two-year-old read words as big as hippopotamus and so forth. I jumped up screaming, "Oh my God!" my baby can read. While I was away, Luck was steadily building Dooney's educational foundation. He is constantly working with him on different assignments until this day, and he makes him think outside the box. We discuss current events, African American pioneers, sports, money management, and basic life skills. To some people, all of this may be typical, that's what you are supposed to do, but when you are witnessing it firsthand, it's breathtaking. I couldn't walk into a "Build-a-Man" store and put something together as perfect as God has created my Boaz for me. He had literally changed my life for the better, uplifted me when I was down, continuously sweep me off my feet, and challenged me to be great every day. Many women look at a woman as speaking highly of her man as if she has to be making this up. Every man isn't the same, and all men aren't boys. No matter how good a relationship is, it still takes work and commitment. You have to commit to love, honor, respect, and value your mate. I am blessed that my husband took a chance on this smart mouth, broken little girl from the hood. Our Williams household family motto is "we didn't wake up to be mediocre." We will rise to set the standard and not conform to the standard someone else sets for us. We are perfectly imperfect for each other while trying to be the best parents that we can be.

From Green to Blue

In 2004, I left the Army and joined the Air Force National Guard. I was hesitant but opened to the new challenge that lied ahead. I was introduced to the recruiting office supervisor, MSgt Vonsetta Roberts (Love). She was prior service with the Army and served time during Operation Desert Storm. MSgt Love was able to speak to me in Army terms, which made my transition into the Air Force a lot easier. I was a sergeant (SSG)/E-5 at this time, so leadership and followership were ingrained. I felt like I could give the Air Force a shot. Once my transition was complete, MSgt Love took me under her wing and became my mentor and trusted friend. She answered all the crazy questions I had and volunteered me for EVERYTHING. I simply rolled with the flow and accepting every challenge she signed me up for. She told me she saw something in me. I was squared away and very mature for my age. She would always say, "Taya, you are wise beyond your years." That really made an impact on me, and I definitely didn't want to let her down. I sometimes overheard her talking about me to different commanders while singing my praises. She always believed in my abilities and challenged me to be my very best. I knew I couldn't let her down, so I made a point to represent her well at every opportunity I had. My admiration for her stretched far beyond what she knew at the time, but as the years went by, I often told her how much I valued her. My mom was extremely grateful for the impact MSgt Love made on my career, and she often expressed her gratitude to her. Our relationships quickly evolved from mentor and mentee to sister and friend. I met several other influential women who embraced me as big sisters to help me navigate throughout my Air Force career to include Vivian Kelly Scott, Cynthia Maull Jackson, Lillie Wilson, Lisa McDay, and Marydee Hidle Wisdom. In every aspect of our lives, we need a village. I'm proud to say through it all it was clear we needed to become our own village to thrive.

As I embarked on my journey with the Air National Guard, things were off to a great start. I was immediately informed of how bad the climate was based on racial issues, and how African Americans were overlooked for good jobs based on "the good ole boy" system. In my heart, I wanted to believe those days were a thing of the past, so I remained optimistic. I was equally afforded opportunities in my direct squadron, which made accepting the reality that the "good ole boy" system still existed difficult. It was not until I applied for a job outside of my squadron that I got my first taste of what the culture was really like. Although I was more than qualified for the position, it was given to a white female with less experience and qualification. Just before the interview, I was approached by a white male, and he stated I should get the job based on my qualifications; however, he emphasized I wouldn't get it because he knew the person who was going to get hired. I asked why he felt that way, and he replied, "This is just the way they do things around here." I did not want to accept it at the time, but he was absolutely correct. I wasn't selected, and the person he was speaking of ended up with the job. Feeling defeated, I went home crying to my husband, Luck.

Guess who was sought after to provide training for the under-qualified candidate. You guessed it, me! While she was supposedly "selected" over me based on experience, I was the one responsible for training her for the job I knew I deserved, the job my superiors clearly knew I could do, and teach. I answered questions and lent a hand way more often than I cared to. One day, I said enough is enough. I went into my commander's office after I pondered over the situation the entire weekend. I told him I needed his help with stopping the phone calls from the chief that selected the individual for the job. I reminded him that the chief stated the sitting candidate was more than qualified for the position than myself and told him the entire ordeal was an insult to me. My commander looked me dead in my eyes and assured me he agreed with and understood where I was coming from, further ensuring he would put a stop to it. I was beyond relieved, not just to have his support, but because I knew as long as they called me directly, I was going to help. That is in my nature and an example of my overall character. At that point, I made a vow to take care of myself, reminding myself that saying no doesn't make me any less of a team player. As time moved forward, I was reluctant to put in for other positions, fearing the same thing would happen. People would approach me and recommend jobs for me to apply for. Regardless of how qualified either of us believed I was, I would say I wasn't interested. The truth of the matter was I didn't want to have egg in my face again if I wasn't selected for a position I qualified for. So many people saw my potential and knew I could get the job done, but that meant nothing when the selecting official didn't see past my skin color.

I didn't know it at the time, but my big blessing was brewing behind the scenes. In 2008, there was an active duty position available that seventeen applicants had already applied for. I had no plans to apply based on history. There were so many people who reached out and told me I "had to" apply for that particular job because they needed me in that office. Again, I was reluctant and brushed it off! After speaking to my mom about it though, she encouraged me to do it. She told me, "Something about this one feels different," and everybody knows that mommas are always right. I decided to put in my application and said, "God, it's in your hands. You know me and my abilities better than anyone, so if it's for me, let your will be done." On the day of my interview, I was number 1 of 18, and I felt the pressure. In preparation for the interview, I put together a visual aid to assist the panel with my vision of the position and how I could assist with taking it to the next level. My nerves were through the roof. It's not every day you get to sit down and interview with the wing/base commander. I felt the sweat rolling down my back, but when it was over, I knew I had given it everything I had. I walked away thinking the next seventeen applicants have one of two choices, either make them forget about me or rise above me. I was notified a few days later I got the job. I was ecstatic and extremely grateful for a wonderful opportunity. Little did I know, this was the beginning of yet another battle.

The micro aggressive attacks from some of the Caucasian females who applied for the job began before I could even get settled into my new position. What stood out the most to me were the comments they made amongst each other. One of the females asked, "Why did they pick her?!" while another was even bold enough to approach the commander and asked him why he picked me? I was troubled by a lot of the attacks, especially when my performance was being questioned. I personally did not question my ability to do the job, but continuing to do so under the scrutiny of others began taking a toll on me. Before things got too far out of hand, I immediately sat down with the commander and expressed my concerns. He quickly reassured me he was pleased with my performance, and he was certain he chose the correct candidate for the position. After that conversation, I was no longer concerned about the naysayers and focused on my job and the mission while implementing much- needed programs and policies. Soon after, I was promoted to master sergeant (MSgt)/E-7 in 2009 by the same commander for my efforts and dedication. Throughout this time, there were multiple deployments and temporary duty tours; however, it was all worth it.

Another memorable deployment was Balad, Iraq, 2009. MSgt (Retired) Kimberly Jones and I were battle buddies, and we literally stayed close to each other the entire time. We were under constant attack, and the sirens were going off quite frequently. This mission didn't require us to go outside the wire, but we were told that our base was a main target, so the locals were doing their best to take us out. Kim and I had finally gotten off work and changed into our physical training clothes and decided to catch the chow hall before it closed to get some decent food. Deployment food can be the worst, boring, and repetitive. It's not like we could run down the street to our favorite restaurant, but we sat up many nights talking about all of the food we were going to eat when we got back stateside.

We reached the chow hall, cleared out our weapons in the barrel, and proceeded to the serving line. When you clear out your weapon, you are ensuring there is no round in the chamber, and the weapon firing selector is placed in the safety position. During our usual small talk, all of a sudden, we hear a loud boom. The chow hall started to shake, alarms started sounding, and we heard the loud voice system alerting us we were experiencing incoming bombs. We were under the closest table we could find. The chow hall was covered with crawling bodies all over the floor looking for cover. Everyone was crying, screaming, and praying. I looked for Kim, and she was nowhere to be found. I panicked. I only wanted to find my friend. When we finally received the "all clear," I found her. We hugged and cried and ran back to our living compound. We were silent in our room as we just sat there and tried to grasp what just happened. We counted our blessings, prayed, and talked about our families for the rest of the night. Sleep was hard to come by after that experience as the days went by.

As if being in hostile territory wasn't enough, I had an encounter with a senior noncommissioned officer (SNCO) who kept flirting with me to the point where it became extremely unprofessional and downright unethical. He showed up to our room knocking, but we never opened the door. No matter how many times I asked him to stop persistently pursuing me, he continued. Many other SNCOs witnessed him saying inappropriate things to me, but instead of correcting him, they laughed it off, as if it was okay. I began going out of my way to avoid him after my duty day. Once my duty day was done, I spent most of my time in my room and went to the gym at a different time than normal. I even cut off any communication during the duty day. If I did not have a professional reason to speak with him, I didn't. Kim and I had a signal for when he was around. If he was near the vicinity, one of us would discretely notify the other, and we would go in the opposite direction. Other than avoiding the SNCO and praying we did not have to experience another attack, our days were spent counting down the time until we reached American soil to be reunited with our families again. At this point we were clearly over the deployment from the excessive heat waves, massive sandstorms, and constant base attacks.

My normal routine when I arrived at work was to log into my computer and began checking my e-mail. One morning in July 2014, would be no different. I came across one e-mail from a sergeant who I ran into the day before. When we ran into each other, we had small talk, got caught up, and went our separate ways. The e-mail message was marked urgent, so I opened it immediately. The sergeant stated he needed help with something very important and asked me to call him as soon as possible. Without delay, I called him on his work number and told him I received his message to call. He told me he didn't want to talk on the work phone about the important matter and asked if I could call him from my cell to his cell? I figured he wanted to discuss retirement paperwork assistance since that was the gist of our conversation the day before, so I obliged. He gave me his private cell, and I hung up the work phone and called him from my cell. He answered the phone on the first ring, and I said, "Hey, what do you need help with?" Everyone knows I will help anyone I can, military or civilian. He proceeded to tell me he had never been with a black woman, and it was a dream of his. Then he told me he wanted to know if he could perform oral sex on me. I sat there with the phone up to my ear, mouth opened, and speechless. I was flabbergasted as I responded, "Are you serious! Are you really serious right now?" He responded, "If you don't want to do it, you can find me another black female." He went on to say he would pay us and allow us to do him back in return. I completely lost it on the phone with him! I told him how disgusting

he was, that he wasn't fit to wear the uniform. I asked him, "Who gave you the damn authority to call me with this sick shit!" I told him, "You need some serious help if you think you can approach me like that, you sick piece of shit!"

After hanging up, I was stuck in a trance. My mind was running rampant, hands shaking, and heart racing. I replayed our conversation over in my head as I wondered what I had said to him the day before to make him feel like this type of disrespect was okay or even warranted. I started blaming myself for literally engaging in a friendly conversation with him. After the shock wore off, I notified our sexual assault response coordinator and our equal opportunity officer. By this time, I was a five-year certified victim's advocate, so I knew the steps to take this time around. As I explained what had just happened to me, I became enraged and literally felt like he didn't think what he said to me was wrong. I started thinking about all the other women he felt this comfortable with or used his rank/position to get younger airmen to submit to horrific requests. This situation was elevated to the highest level. As the situation was being investigated, this bastard was still texting asking if I made up my mind or if I had found someone else willing to do it. The only reason I didn't block him was because of the investigation; I used all of his communication as evidence. I was on high alert while things were being handled. I saw him everywhere I went and felt like he was intentionally showing up just for me to see him. At the end of the investigation, he was found guilty, and the proper reprimands were issued for his behavior. I have always wondered did he ever have any remorse for disrespecting a black female or just sorry he picked the wrong one to approach with such vile disrespect.

Daddy's Girl

For me, my dad was the in-house superhero who didn't wear a cape. Daddy wore steeled toed boots, a tool belt, hard hat, and often came home from work covered in paint, and carrying a ladder. Those were the capes he wore Monday through Saturday, and on Sunday, he wore a nice suit and Stacy Adams shoes to church. My dad showered me with love, life lessons, and the gift of Jesus. My dad established the understanding we would attend church before we did anything else on Sunday. At any given moment, you could expect to have a heart to heart talk with him because he was intentional in always knowing what we were doing and how we were doing mentally and emotionally. I hated to do that back then because I felt like I would have to bare my soul to my dad. In reality, we were building a strong bond I didn't appreciate until I was old enough to understand.

Growing up, I was my daddy's shadow. He took me everywhere he went, and I loved it. It seemed as though everyone knew of him; therefore, everyone knew me because I was his daughter. I couldn't go anywhere without someone saying, "Hey, you are Scott's daughter, right?" I'd smile and respond proudly saying, "Yep, that's my daddy." I literally believed he was just *my* daddy, even though I had brothers and sisters older than me. I didn't think much about that at all because he was simply *my* strong, smart, humble, hardworking, and loving daddy. The one who made sure we were well taken care of and loved. My dad had analogies for *everything*. No matter what happened in our lives, he truly had a way of giving advice with clear understanding. It seemed as though he always knew when I was having a bad day or simply just needed to talk. He'd come home, and I would run up to him, hug him, and he'd give me a kiss and say, "Tell me all about it." He listened extensively without interruption, and more importantly, he never judged me.

I remember when I was about five years old, we started this thing where he would ask me how much I loved him, and I would stretch my arms open wide and say, "I love you this big." One day, while wrapping up a week-long military assignment in Washington, D.C., while in the air terminal after I had my bags checked, the rental car turned in, and found out our departure time, the commander was briefing us on the success of the trip and mid-sentence my phone started to ring. I looked at it, and it was my brother Prime. I wasn't surprised it was him because he always called and checked on me. I figured I would call him back after the commander's brief. Right after the brief, he was calling me back, and I told him I was about to call him back. He said, "Where are you now?" I said, "In the terminal, preparing to step onto the plane." This time though, Prime's voice was different, and I could tell something was wrong! He finally told me my daddy was unresponsive, and he was in the hospital. As soon as we landed, I called Prime back and was told my dad was still unresponsive. He said my dad was having mini seizures, and they couldn't determine why. Upon touching down in Montgomery, I went directly to the hospital without delay. Daddy had been diagnosed with Type 2 diabetes and high blood pressure many years back. A few years before this incident, he had been diagnosed with onset dementia to the point where I recognized he was sometimes struggling. He began forgetting certain things and would repeat himself quite frequently during our phone conversations. Later that evening, the hospital told us they were going to send daddy to Emory hospital in Atlanta so that a team of specialists would be able to take over his care. My cousin Vickie and I headed out for Atlanta the same night, praying and crying the entire way. My mom, my siblings and I were at the hospital, hoping to get a clear understanding of what was going on with daddy. After a week of testing, we were finally informed the seizures ruined his left side of his brain, and he would never be able to live without the ventilator and would remain in a vegetative state. We had to make one of the hardest decisions of our lives, which was to remove the ventilator and place him in hospice care. I was devastated and couldn't believe we were faced with that decision. That was the very first big blow to our immediate family. I signed the papers for his release and stayed in the room with him, talking to him and singing some of his favorite songs. I prayed for a miracle; however, I knew we were about to face the inevitable soon. I watched him all day as I listened to him struggling to breathe on his own. It was heartbreaking, to say the least, but I was hanging on to the last few moments God allowed me to have with him. I couldn't sleep that night because I was listening for him to tell me he was okay. I finally drifted off to sleep, and at that moment, he took his last breath. I jumped up and ran over to his bed as I called for the nurse. Finally, the on-call doctor came in and pronounced him dead at 0100 May 17, 2014. I was numb! I started to panic and hyperventilate because even though I knew it was going to happen, I was still not ready to lose *my* daddy. I beat myself up because the day before he became ill, I had intentions of calling him, but I decided against it. I looked at the phone, and due to the time difference in D.C., I said I would wait until tomorrow, but tomorrow for me was too late.

I struggled for a while to forgive myself for that.

As time went on, I took comfort in knowing Daddy knew I loved him. My mom was our strength and anchor during this tough time. She gracefully walked with us, her children, every step during the grieving process. I have so many fond memories of him, and they will always be kept hidden in my heart. I will forever love you, "this big, Daddy."

My Ruthie, My Heart

I left for a six-month deployment on my birthday, July 12, 2018. While settling into my temporary desert lifestyle, I still had the chance to FaceTime my mother daily. My husband, Luck, went by my mother's house every day on his lunch break to ensure I would get to see them both at the same time. My mom was by far the best mother a girl could ever wish for. She and I were the perfect "mother-daughter duo." My mom went out of her way for all five of her children, her grandchildren, and great grands. She was a hard worker, a great cook, and the family comedian who loved to dance. My Ruthie was truly the life of any party, and she believed everyone's birthday celebration was hers.

My mom and I talked about everything. As a child, I always thought she had superpowers. She knew where everything was that was lost or misplaced, and she had a solution for any problem. My mom was a great example of what faith looked like. She was honest, strong, and was the epitome of love. Like she had done on all my other deployments, my mom wrote me letters and sent a card to me daily. However, there was something different about her most recent cards and letters. We were always sure to express our love for one another and ended our calls with an affectionate, "I love you".

During this six-month deployment, we had phone calls where we laughed so hard we couldn't catch our breaths, and we had mushy conversations sharing our mother-daughter affections toward one another. I even had the privilege of sharing this book you are reading now with her. I shared the vision that I had for "Girl Pals." I told her what the colors symbolize and why I wanted to create something for women empowerment. My mom told me, "Taya, you can do anything you put your mind to because you are a smart young lady, and I am blessed to be your mother." I responded by reminding her of something I've always said, "No, Mom, I am blessed to have you." This was our favorite phrase to each other. We were literally best friends! I was definitely my momma and daddy's baby; everyone knew that about me. My deployment was coming to an end, and my mom had planned my welcome home party to the tiniest detail. She even began to prepare my favorite foods because that was the type of mom I had. We talked about some of the things we were going to do once I got back, especially playing cards at her house on Sundays, which was mandated family time. I couldn't wait to get back to that table and catch up, laugh, sing our songs, eat some of her good cooking, and tease her about some of the crazy things she wrote in my letters. We had a lot to catch up on as she would say over the phone, and she couldn't wait to kiss my little lips when I got off the plane. She also told me she was going to be screaming the loudest at the airport when she could see me walk through the gate. I often visualized our embrace with one another as she said things to me to make me anticipate my homecoming.

I found out that my return date to the states was officially January 19, 2019. Elated, I called my husband, and he, in turn, called my mom to inform her of the concrete date. When my husband called my mom, she told him that was the best news she heard all day. On FaceTime, my momma showed me the calendar she had been counting the days down on, and she would always say, "We don't have long Taya." I would just laugh and say, "You are so crazy, Mom." I was at my desk getting ready to end my duty day on January 16, when my husband FaceTimed me. The phone call caught me by surprise because we were nine hours ahead, so it was pretty late central standard time. I answered the call, and before I knew it, my world literally came crashing down. Luck informed me my mom had unexpectedly had a heart attack and passed away. I was stuck! I felt every emotion you could feel at one time, and I couldn't understand what had happened. Momma and I had just talked, and now this? I was literally three days closer to being reunited with her, where we would do all the fun things we talked about, shopping and partying, but what in the hell had just happened? I was in a daze until I was flown out to the states to be with my family. The meeting at the airport wasn't what we planned. It was the saddest time I have ever had upon reconnecting with my family. Instead of her screaming the loudest because I had finally arrived, I was screaming the loudest because she wasn't there. This just didn't seem right! I was waiting for someone to tell me they made a mistake and it wasn't true. The level of pain I felt during the passing of my beloved mother was a type of pain I had never experienced, and it sunk me to the bottom of the floor. My Ruthie's final chapter would be an invisible scar that lead me to my biggest breakdown.

Invisible Scars

For over a year since my mom's passing, I have suffered from depression and anxiety, and I am currently finding ways to cope and get back to myself. I have days where I literally don't want to get out of bed, experience poor appetite, sadness, and prefer isolation. I also had to brave the world to pretend I was okay, while deep down inside, my heart was shattered. I was literally walking around in a daze with a thousand things running through my head. I had to wear a "mommy is okay" face for my son, who I didn't want to see me so down. Unfortunately, he walked in on me in the middle of me having a "moment" several times. I was so lost in my own pain I couldn't see that my son was hurting just as bad from the loss of his second mom as I was and had a greater fear of losing me. I can't even begin to describe the feelings I have to anyone that would remotely make them understand the hollow in my heart. My mind couldn't or wouldn't let me understand how something so tragic could happen when my mom and I had so much catching up to do.

Memories were everywhere, and I couldn't escape them no matter what I did. I could hear a song in the car and cry, watch a commercial on television and cry, pick up my phone to call her and realize I couldn't and cry, go to her house to realize she will never be there again and cry, remember a funny story and cry, remember to tell her something I forgot to tell her and cry. I needed her so badly for advice on how to deal with the pain of losing her, but she didn't leave instructions. All my days consisted of crying and feeling empty. My prayers felt like they weren't getting through, just like my attempted phone calls to her. I realized I couldn't call my friend like I did every day. I couldn't stop by to see her on my way home from work like I did every day. I couldn't spend my weekend with her like I did every weekend. No matter what my husband and I planned, our last stop would be down to Ruthie's house. She looked forward to us coming just as much as we looked forward to going. That was our home away from home, where she made us feel the most welcomed and hated to see us leave. It was nothing but love that surrounded us at her house. How does all of that just vanish and disappear? I couldn't understand what happened. I definitely couldn't adjust to the new normal my life was experiencing. How hurtful is it to be away on military duty and come back home to the loss of the one girl in the world who loves you the most? In this loss, I lost a mother, friend, protector, prayer warrior, motivator, nurse, doctor, teacher, counselor, and coach.

I have been blessed with the best team of doctors and therapists that are committed to my wellness. I have faith, and I pray as well, but I am also human. Many things I have experienced during this journey are things I do not wish upon anyone. I didn't even recognize who I was, and I still have my struggles. I am blessed with the best circle of support imaginable, and I know God has his hands upon me; and the testimony I have to come will be one for the masses. I know this low place I am slowly climbing myself out of is not my destiny. I view life, people, and relationships differently than I did before, and I am on my way to living in victory.

There are so many judgmental people who are so misinformed and lack education surrounding depression. The lack of knowledge has deterred many from seeking the help they so desperately need. Most folks depend solely on religion to overcome their obstacles, but God created therapy along with faith and prayer to assist in recovery. Most are afraid to open up and ask for help a lot of times in fear of what they disclose or confide is later used against them to hurt them. The world has replaced compassion, empathy, and sympathy with malice and hatred. If you know you are not totally in the fight to assist a brother or sister in need, please do not step in the ring, pretending to assist. You will only cause more harm than good. This is a serious issue plaguing all communities, and no one is exempt. Your social status, occupation, education level, or wealth doesn't exclude you. People of faith forget to be human and judge your faith when you don't react the way the church thinks you should when you profess to know God. Instead of faith bashing, use your strength, faith, and prayer to help pull the individual along the way in love and understanding.

I know my depression comes from the way my mom loved me as a child and even more into my adult life. I am strong, but I was still her baby. She was my voice at times when I couldn't speak, she was my eyes when I couldn't see things for what they really were, she was my strength when I was weak during troubled times in my life, she believed in me when others doubted me and caused me to doubt myself, and she always saw the best in me no matter the situation. I know now not to judge any situation I haven't had the experience in dealing with because you can't speak from the spectator box when you literally have no idea what a person is truly going through. I haven't had a good night's sleep in a year, but I can happily say my counseling and medicines are helping me in moving toward getting to the other side of healing and wholeness.

Grief is definitely a process, and I have

experienced all phases of it, plus some. Things have literally changed since she's been gone, but I will carry her in my heart forever. I lost my dad in May 2014, and my mom January 2019, now both of my parents are living in heaven. Many days I feel like a parentless lost child. I am sharing my struggle with transparency and vulnerability because I know this test is setting me up for a fire burning testimony. I know one day I will be able to tell my full story of how I made it over, but I am hoping this snippet encourages another Mataya in the world who may be dealing with this type of loss. I can assure you we will get better! I am going to keep on moving forward until I reach my higher ground.

My Final Stripe

July 1, 2019, I was promoted to chief master sergeant (CMSgt)/E-9. Taking care of the people I have charge over is still my priority and will remain even into my transition into the civilian sector. I am extremely blessed and fortunate to have made it to this level in spite of some of the things I have dealt with throughout my career and personal life. Everything I experienced that was negative didn't need to be shared, and I didn't want to disclose too much because through it all, I am surviving. My God has blessed me with the best leadership to end my career under, and I am grateful. Many people will come up against you not because of how bad you are, but because they know how great you can become. When your heart is pure, and intentions are well, the Lord will ensure your good name remains intact for the right people. Those who speak ill of you and spew hate are only showing who they really are deep down inside. Shine on! God has truly blessed this young black girl from the projects. I have more than I ever thought I could achieve. We can't accept the labels and tags that other people place upon us. Our God created opportunities for us ALL, but men have put stipulations on us because we look different and speak differently. I have many more goals I aspire to reach, and I claim every great thing the Father has in store for me. It's hard to celebrate yourself because an insecure person would view you as bragging or boasting. The boast goes to God for allowing us to eat the good of the land. We have not because we ask not and won't put in the work to achieve our goals. I am striving to change the lives of the people God has entrusted me to lead.

We all know that if there is no vision, the people will surely perish. I am trying to secure my future and plan for my life outside of the military. Soon and very soon, these boots and fatigues will be traded in for civilian clothes. I have big ideas, and I just want to build the table that many can eat from and turn that table into many more for the generations to come. We lose sight of the end goals when we allow pride to enter our life. This nasty trait will have you hating a person who possibly could give you the next boost to jumpstart your business or assist in building your brand. Do we really want to work together? Or just talk about working together on our posts and memes? United we stand but divided we will *never* rise. Let's fix each other's crowns and get to work. America needs us to be the best version of ourselves as we spread love, peace, wisdom, and knowledge. Many times, I wanted to give up and just throw in the towel because there were so many things I couldn't understand. God would always throw it back and say you will see this journey to the end. I am extremely grateful he cares for me and truly knows my heart for people even when others may misunderstand my good intentions, judge my words or deeds, and find fault in what they truly do not understand about me. My God knows my heart, and he definitely knows me through all of my flaws and shortcomings. I am still learning, trying, growing, and endeavoring to be all he created me to be (Jeremiah 1:5). Throughout our journey in this life, no matter what path we choose, we must stay in power and prayer, and we will win—one love.

In that same year, July 27, 2019, I lost one of my favorite sister-cousins, Vickie McKenzie. I started to decline and sink back into the place from where I had begun to make minimal progression. It was like someone set the emotional reset button, and I was emotionally drowning all over again. The sad part is she was very instrumental in my struggle after I lost my mom. Now I lose you too?! We had a special bond I will cherish for the rest of my life. My Auntie Stella, her mom, called us Frick and Frack or whatever label was used to describe our bond; we were definitely more than cousins. Our personal talks about life issues, me telling her about all of my bright ideas while she boosted me up, the crazy hysterical laughs, our girls' nights while she boosted my Pistol Pete alter ego, her being my personal hype man when I danced, her making me believe I could sing when I lead solos at church, and our genuine mutual support for one another will never be forgotten. Her acts of kindness and support for me alone were genuine and unmatched. I never met someone who wanted something so bad for someone else without looking for anything in return. She was big on family and loved us all, no matter what went through.

We had a long conversation the day before she died. She called to give me my safety briefing and demanded I enjoyed my trip. The next day my husband and I were heading to Jamaica for our fifteenth wedding anniversary celebration. She told me she knew I really needed the trip, and I better enjoy myself no matter what. She said, "I am serious Taya, you better not think about anything back home, and I mean it." I told her I was excited and was really looking forward to spending some time with Luck and having a few good drinks. She said to me, "God told me to tell you, you will always be blessed because of your heart. You give unselfishly and look for nothing in return. All we have to do is call on you, and you will be there for anyone, and I am blessed to call you my sister-cousin." I told her to stop all of that before she made me cry! She said, "My Taya is a chief, and your mom and dad are so proud of you, and they are looking down on you from heaven." I told her how I wished they could be here to see their baby, and she said, “Don’t worry, they see you...trust me, they do.”

We talked on the phone for over an hour, and she said I am going to let you go, but I will text you in the morning. She told me to take plenty of pictures, she loved me and was super proud of me. She told me she had her scriptures ready and printed out for our cousin Cedric’s funeral, and she was ready. She said, “You did a great job coordinating everything, and I love seeing how serious you get when you get in your zone. You get focused and handle your business and make everything happen; that’s your gift.” I told her how grateful I was for her, and I loved her more than she knew. She said, “Oh, I know you love me, Taya.” That was the gist of our last and final phone call. I have replayed it over and over in my head, looking for something to tell me that conversation was the final one. Her love for me is irreplaceable and is truly missed. She was a shoulder for me to lean on when I lost my mom, which Vickie and my mom also had a wonderful bond. She was taking my mom’s passing pretty hard too, while still sending her daily text messages to me and my brothers and sisters. We talked, texted, and prayed daily up until the day she died. If I didn’t answer the phone when Vickie called me, she knew I was having a moment and would show up at my door to check on me. If I would be sitting at my desk in a full-fledged meltdown, she’d out the blue send a scripture or something encouraging. She truly understood me and was always willing to help me with whatever projects I was working on because all she wanted to do was to help me. Every now and then I would surprise her with something just to say thank you, and she would just cry because she and I both were crybabies. We would go out to eat and just laugh and joke about everything.

College football season will never be the same. She was a diehard Roll Tide fan and declared herself a coach Nick Saban, "Nick chick," and would pull out her phone pretending to call him as if he knew her personally. We would be on the phone watching the game, and you would think she and I were both coaches the way we would call plays or suggest what "we" should do next as if we were a part of Alabama Crimson Tide staff. I still read my last two cards she gave me right before she passed away. I have re-read our text messages, WhatsApp messages while I was deployed, and have gone through the million and one pictures of us in my phone. We had two major celebrations in the month of July 2019. I got promoted to chief master sergeant, and we celebrated at karaoke; my birthday was celebrated two weeks after that. We went out, and she hyped me up to ride the mechanical bull. Of course, I did it because she was truly my hype man. She believed in me when I didn't believe in myself at times. Whatever it was, she could pray and encourage me through. It's been a long day without you, my cousin, sister, and friend, but I will tell you all about it when I see you again. I am convinced that her, my mom and dad are truly getting on Jesus' nerves concerning me. Some people you just cannot forget about or even replace. I will love them all forever and a day. It seems like heaven just couldn't wait for them.

Thank You

Family—Thank you to my sister Sophia, her husband, Greg, my cousin Tiffany, and her husband, Robby, for being amazing role models in the military and in marriage. It's refreshing to see true love still exists. Many thanks to my sisters, brothers, nieces, and nephews for *always* being the *best* part of my memories. The love between us all is beyond anything I could have ever hoped for or even imagined. There are so many families in turmoil and chaos, but God is holding us together in the absence of our foundation, parents and grandparents. I am so thankful we were raised in love, and when we had nothing left, the love remained the center point of our existence. I am proud to be a part of this lineage, and I know the love and support amongst us all will outweigh any bad day or the next storm. God was truly careful when he created this bond that has banded us for life. We must continue to hold fast to the promise and continue to do our work. Even with our backs up against the wall, God has continuously been our refuge. I love all of you *amazing* people.

Husband and Son—Luck and Dooney, there's definitely no me without you, and no we without us. I am the better version of me, because of the two of you. This love thing is so real! Thank you both for your unwavering support and encouragement. As we talk about our table, God is the first leg, babe you are definitely holding down the second leg, you got me holding down the third leg, and Dooney holding down the fourth. We know God is our anchor, but in order for us not to wobble, we three must continue to do our part. I will love y'all forever, WilliamsFam3.

My Sisters of Love: #GirlPals—I have never met a group of women with so much love and support in their hearts than the six of you. You all have been an important piece in my life. I have drawn different things from each of you at different stages in my life. Each of you brings the true meaning of sisterhood and friendship to our relationship. My bond with each of you is twenty-plus years strong, and I cannot thank you all enough for your consistency, realness, maturity, loyalty, honesty, and love. You all have changed the narrative of where they say women don't know how to get along, "tuh" I guess the ones saying it must have all the wrong friends. I wanted to group you to show you I value each of you equally, and I love you all immensely. There is something uniquely special about you, ladies. It amazes me every time I think about how all of *my* friends became friends—that's dope! We've laughed together, cried together, encouraged each other, disagreed and resolved amicably, partied together, supported each other's kids, prayed together, prayed for each other, and everything else that true girlfriends do. Whenever we fall out, it's from laughing so hard until our belly aches. Thank you all for the love and support throughout the years, and here's to twenty-plus more years of fun. #GirlPals: Missy Wright, Evita Williams, Katrina Greenwood-Williams, Monique Randall, Jameca Barnes, and Kimberly Jones.

Made in the USA
Monee, IL
05 November 2026

67947703R00049